The Ethical Steward
Sustainable Deer
Management Practices

"The care of deer reflects the ethics of the steward; their thriving, a testament to the harmony between land and keeper."

Benjamin.
Deer Manager

This guide is designed not as a comprehensive solution or an answer to every deer land management issue, but as a starting point for those seeking to address their specific needs in this domain. Having devoted just over two decades to deer management, I continue to find there's much to learn. In a modern landscape where time constraints and the quest for immediate results prevail, there's a concern that traditional land management teachings might be overshadowed by digital and technological advancements. It's my earnest hope that across the UK, seasoned deer managers will continue to impart their advice to those willing to learn, ensuring that the industry remains a valuable resource for future generations, just as it has been for us. Through shared knowledge and a dedication to the principles of land management, the essence of this profession can continue to flourish amidst evolving modern challenges.

Contents

FOREWORD

The realm of land management is a multifaceted domain, demanding a thorough understanding and effective management of the diverse ecosystems residing within the land. Among the residents of these ecosystems, deer have a significant presence, whose management necessitates a balanced approach to ensure the sustainable co-existence of wildlife and human activities.

"The Ethical Steward: Sustainable Deer Management Practices" is a resource aimed at addressing the practical and basic aspects of deer management within the United Kingdom. This guide provides a structured approach towards understanding the biological, legal, and societal frameworks surrounding deer management, particularly from the perspective of landowners and managers.

We have systematically explored the necessary facets of deer management, integrating scientific knowledge with practical strategies. The narrative begins with an introduction to deer management, explaining its importance, objectives, and the legal frameworks guiding it. It then delves into the biology and behaviour of deer, providing a foundation for the subsequent discussions on population monitoring, habitat management, and population control strategies.

The guide also addresses the critical aspects of human-deer interactions, health and disease management, and the economic considerations surrounding deer management. These sections are designed to provide landowners with a well-rounded understanding of the challenges and opportunities present in deer management.

The significance of community engagement and education is highlighted, promoting a collaborative approach towards sustainable management. The narrative emphasizes the importance of adhering to policy, legislation, and compliance, which are crucial for implementing effective deer management practices on a broader scale.

In summary, this guide has been put together to support landowners and managers in navigating the complex terrain of deer management, promoting a harmonious balance between human activities and wildlife conservation.

Benjamin
Wildscape Deer Management

CHAPTER ONE

INTRODUCTION TO DEER MANAGEMENT

IMPORTANCE OF DEER MANAGEMENT

The practice of deer management resides at an intersection of ecological stewardship and societal responsibility. Its importance is varied, encompassing impacts on biodiversity, habitat integrity, and the dynamics of human-wildlife interactions. Understanding the importance of deer management requires a deep dive into these various domains, shedding light on the interconnectedness of deer populations with broader ecological and societal frameworks.

ECOLOGICAL IMPORTANCE

Biodiversity Maintenance: Deer, as integral components of their ecosystems, play a pivotal role in biodiversity maintenance. They serve as both browsers and grazers, influencing vegetation structure and composition. Their browsing habits can create more diverse habitats by preventing any single plant species from becoming dominant. However, when deer populations are uncontrolled, over-browsing can lead to the decline of certain plant species, disrupting the ecological balance and negatively impacting other wildlife species that share the same habitat.

Habitat Integrity: Deer populations significantly influence the health and diversity of various habitats. Through their foraging habits, deer aid in seed dispersal, which is crucial for plant regeneration and habitat restoration. However, overpopulation of deer can lead to overgrazing, which not only degrades the habitat but also leads to soil erosion and water runoff, further exacerbating the loss of habitat integrity. Effective deer management ensures that deer populations are kept within sustainable limits, thereby promoting habitat health and resilience.

SOCIETAL IMPORTANCE

Human-Deer Conflicts: Deer populations intersect with human communities in ways that can sometimes lead to conflicts. The most common forms of human-deer conflicts include deer-vehicle collisions, which pose significant safety risks, and agricultural damage, which can have economic repercussions for farmers. In urban and semi-urban areas, deer can venture into gardens and landscapes, causing damage and creating nuisance. Effective deer management practices are crucial in minimizing these conflicts, thereby safeguarding both human and deer populations.

Economic Value: Deer also hold economic value, which is realized through venison production and recreational hunting, both of which contribute to local economies. Well-managed deer populations support sustainable hunting practices, which in turn can generate revenue for local communities and conservation efforts. Viewing deer in their natural habitat can also be a source of eco-tourism, further contributing to the economic value of deer.

Educational and Cultural Value: Deer have been emblematic in various cultures and hold educational value in terms of wildlife appreciation and ecological education. They serve as a gateway for individuals, especially children, to connect with nature, fostering a sense of stewardship towards wildlife and natural habitats.

The endeavour of deer management is thus a nuanced and significant task, with ramifications that ripple through ecological and societal spheres. Through effective deer management, a harmonious balance between deer populations, their natural habitats, and human communities can be achieved, fostering a sustainable coexistence that upholds biodiversity, habitat integrity, and societal well-being.

OBJECTIVES OF DEER MANAGEMENT

The practice of deer management encapsulates a broad spectrum of objectives, each geared towards ensuring a harmonious balance between deer populations, their habitats, and the human communities they interact with. The endeavour towards achieving these objectives requires a thorough understanding, strategic planning, and effective implementation of various management measures. Here we explore the primary goals of deer management, explaining the significance of each.

Population Control: Maintaining deer populations within ecologically sustainable levels is fundamental to any deer management program. A balanced deer population is instrumental in ensuring the health and vitality of individuals within the population, minimizing disease outbreaks and promoting genetic diversity. Population control measures are employed to prevent overpopulation, which can lead to overgrazing, habitat degradation, and increased human-deer conflicts. Various strategies, including culling, fertility control, and relocations, can be deployed to manage population numbers. The objective is to establish a population size that is in harmony with the carrying capacity of the habitat, ensuring the long-term sustainability of both the deer population and the ecosystem.

Population control contributes to the mitigation of negative interactions between deer and other wildlife species. It helps in reducing competition for food resources and habitat, which is crucial for the preservation of biodiversity. Effective population control requires a robust understanding of deer biology, behaviour, and the ecological dynamics of the habitats they occupy. This understanding facilitates the development of management strategies that are both effective and humane.

Habitat Preservation: Habitat preservation is intrinsically linked to the objective of population control. A well-preserved habitat provides the necessary resources for deer and other wildlife species to thrive. The management of deer populations contributes to the maintenance of vegetation structure and composition, which in turn supports a diverse array of wildlife species. Habitat preservation also involves the restoration of degraded areas, which may include measures such as reforestation, invasive species control, and erosion control.

Habitat preservation is not only beneficial for wildlife but also for human communities, as healthy ecosystems provide a plethora of ecological services including water purification, soil fertility, and carbon sequestration.

Reducing Human-Deer Conflicts: Human-deer conflicts arise in various forms, including deer-vehicle collisions, agricultural damage, and the transmission of diseases. Reducing these conflicts is crucial for promoting coexistence between deer and human communities. Effective deer management strategies aim to mitigate the risks associated with deer-human interactions, safeguarding both the deer populations and the human communities they interact with.

Measures to reduce human-deer conflicts may include the implementation of fencing and other deterrent systems to protect agricultural lands, the establishment of safe crossing structures to prevent deer-vehicle collisions, and public education campaigns to raise awareness about deer behaviour and safety measures. Additionally, in areas where deer populations are dense, controlled hunting or culling may be employed to reduce the risks of conflicts.

DISTINGUISHING BETWEEN PROFESSIONAL MANAGEMENT AND RECREATIONAL DEER STALKING

Management encompasses a variety of practices, each with its own set of objectives and methodologies. Two such practices that often get conflated yet serve distinctly different purposes are deer management and recreational deer stalking. Understanding the differences between these two practices is crucial for estate owners, land managers, and hunting enthusiasts to ensure that deer populations are managed sustainably and ethically.

Primary Objectives:

Deer Management: The primary objective of deer management is to maintain a balanced and healthy deer population within a particular habitat. This involves controlling population numbers to prevent overgrazing and habitat degradation, maintaining a balanced age and sex ratio within the population, and monitoring for diseases.

Recreational Deer Stalking: Recreational deer stalking, on the other hand, is primarily pursued for sport, personal enjoyment, or the acquisition of venison. While it may contribute to population control, its primary goal is not geared towards long-term ecological balance or deer health.

Methodological Approaches:

Deer Management: Deer management necessitates a scientific and systematic approach. This involves population assessments, habitat evaluations, and the development of management plans that dictate when, where, and which deer should be culled. The execution of these plans requires a high level of expertise and often the collaboration with wildlife biologists or other professionals.

Recreational Deer Stalking: Recreational stalking lacks the systematic approach seen in deer management. It is often carried out on an ad-hoc basis, driven by the opportunity and desire of the individual hunter rather than a structured management plan.

Ethical Considerations:

Deer Management:
Ethical considerations in deer management are geared towards the welfare of the deer population and the ecosystem. This includes ensuring humane culling practices, minimizing stress on the deer population, and adhering to legal and best practice guidelines.

Recreational Deer Stalking:
While ethical hunters will adhere to humane and legal hunting practices, the ethical considerations in recreational deer stalking may not extend to the broader ecological impact or the long-term welfare of the deer population.

Impact and Outcome:

Deer Management: The outcomes of deer management are measured in terms of improved habitat quality, balanced deer populations, and overall ecosystem health. The success of deer management practices directly impacts the environment and local biodiversity.

Recreational Deer Stalking: The outcomes of recreational deer stalking are more personal and less quantifiable at an ecological level. Success might be measured in terms of personal satisfaction, skill development, or the acquisition of venison.

The delineation between deer management and recreational deer stalking is fundamental for a holistic understanding of deer hunting practices. While both practices hold a place within the broader hunting sphere, their differing objectives, methodologies, and impacts underscore the need for clear distinction and understanding among practitioners, landowners, and the wider community.

LEGAL FRAMEWORK AND REGULATIONS

The legal framework governing deer management in the UK is an intricate structure designed to ensure the ethical and sustainable management of deer populations. It encompasses various pieces of legislation and is overseen by several regulatory bodies, all aimed at balancing the needs and rights of landowners, the welfare of deer, and the health of the environment. The framework sets the standards for deer management practices, providing guidelines and regulations that dictate how deer populations should be managed across different regions of the UK.

Please note that this is constantly changing therefore it is essential that you check the most up today information by the relevant department.

KEY LEGISLATION

Deer Act 1991: The Deer Act 1991 is a pivotal piece of legislation that governs deer management in England and Wales. It sets out the conditions under which deer can be taken or killed, and outlines the closed seasons for deer to ensure their protection during crucial periods such as breeding or fawning. This legislation also stipulates the requirements for the sale of venison, aiming to ensure that deer populations are managed sustainably and ethically.

Deer (Scotland) Act 1996: This Act is central to deer management in Scotland, setting out the framework for the management and protection of deer in the wild. It establishes the legal framework for deer stalking and culling, and the responsibilities of landowners and managers in ensuring the welfare of deer and the protection of natural habitats from deer-related damages.

Wildlife and Countryside Act 1981: Although not exclusively focused on deer, the Wildlife and Countryside Act 1981 is a broad-reaching legislation that impacts deer management. It sets out the legal framework for the protection of wildlife, including deer, and their habitats, and regulates the introduction of non-native species, which could include certain species of deer.

REGULATORY BODIES

Natural England: Natural England is a significant regulatory body overseeing deer management in England. It provides guidance, licenses, and support for individuals and organizations involved in deer management. Through its regulatory functions, Natural England ensures that deer management practices adhere to the legal framework and promote the conservation of natural habitats and biodiversity.

Scottish Natural Heritage (SNH): SNH plays a similar role in Scotland, providing guidance and oversight for deer management practices. It is responsible for ensuring that deer management contributes to the broader goals of biodiversity conservation and habitat preservation in Scotland.

Forestry England and Forestry and Land Scotland: These bodies are responsible for managing public forest lands in

England and Scotland respectively. They play a crucial role in deer management within these lands, ensuring that deer populations are managed in a way that promotes the health and sustainability of forest ecosystems.

COMPLIANCE AND ENFORCEMENT

Ensuring compliance with the legal framework is critical for effective and ethical deer management. Compliance mechanisms include licensing systems for deer stalking and culling, and reporting requirements for deer taken or killed. Enforcement is carried out by the respective regulatory bodies, and may include inspections, fines, and in severe cases, prosecution for non-compliance. These mechanisms ensure that deer management activities are conducted in line with the legal and ethical standards set out in the governing legislation.

The legal framework surrounding deer management in the UK provides a structured and regulated approach towards ensuring the sustainable management of deer populations. By adhering to this framework, landowners, managers, and practitioners contribute to the broader goals of wildlife conservation and habitat preservation, promoting a harmonious and sustainable coexistence between deer, other wildlife, and human communities.

The Ethical Steward

CHAPTER TWO

UNDERSTANDING DEER BIOLOGY AND BEHAVIOUR

SPECIES OF DEER IN THE UK

The United Kingdom hosts a variety of deer species, each with its unique characteristics and ecological roles. These species vary in size, habitat preferences, and behaviours, and their presence is spread across different regions of the UK. Here we look into the prominent deer species found within the UK, shedding light on their distinctive traits and geographical distribution.

Red Deer (Cervus Elaphus)

The red deer is the largest native deer species in the UK and is known for its majestic appearance. Characterised by its reddish-brown coat and impressive antlers, red deer are commonly found in Scotland, particularly in the Highlands. They also have populations in South West England, East Anglia and the Lake District. Red deer are typically associated with open moorland and mountainous habitats, although they can also be found in woodland areas.

Roe Deer (Capreolus Capreolus)

Roe deer are one of the native species to the UK, and they are smaller in comparison to red deer. They possess a greyish-brown coat that turns to a red or brown colour in summer. Roe deer are widespread across Scotland, northern, and central and southern England. They are adaptable creatures, inhabiting mixed woodland, grassland and urban fringe areas.

Fallow Deer (Dama Dama)
Fallow deer, introduced by the Normans in the 10th century, are medium-sized deer known for their variety of coat colours, which range from black to white, though the common variety is brown with white spots. They are widespread in England and Wales with notable populations in the Forest of Dean, New Forest, and Epping Forest. Fallow deer prefer mixed woodland and open grass areas.

Sika Deer (Cervus Nippon)
Originally from East Asia, Sika deer were introduced to the UK in the 19th century. They are similar in size to fallow deer and have a dark brown to black coat. Sika deer are mostly found in Scotland, but populations also exist in southern England. They inhabit heathland, grassland, and woodland areas, and are known to interbreed with red deer, which can impact the genetic integrity of native red deer populations.

Muntjac Deer (Muntiacus Reevesi)
Muntjac, were introduced to the UK from China in the 20th century. They are small deer with a russet-brown coat and are known for their distinctive barking call. Muntjac deer have a wide distribution across central and southern England, and they prefer dense woodland and scrub areas.

Chinese Water Deer (Hydropotes Inermis)
Chinese water deer were introduced to the UK in the 20th century. They are small, with a characteristic lack of antlers and possess elongated canine teeth. They are primarily found in the fens of Cambridgeshire and along some river valleys in Norfolk and Suffolk. They prefer wet, marshy habitats, hence their name.

These diverse species of deer add to the rich tapestry of wildlife in the UK, each contributing to the ecological dynamics of their respective habitats. Managing these deer species responsibly is imperative to ensure the health of their populations and the ecosystems they inhabit. The understanding of the characteristics and geographical distribution of these deer species forms a foundational aspect of deer management in the UK, aiding in the development of management strategies that are tailored to the needs and behaviours of each species.

DEER LIFECYCLE AND REPRODUCTION:

The lifecycle and reproductive strategies of deer are vital aspects to comprehend for effective deer management. Understanding the phases of a deer's life and its reproductive behaviour aids in making informed decisions in population control and habitat management. Here, we delineate the lifecycle, reproductive strategies, and their influence on population dynamics.

LIFECYCLE

Birth and Early Life: Deer are typically born in late spring or early summer, a period when food is plentiful and the weather is favourable. Fawns or calves are usually born as singles or twins, although in some species and under certain conditions, triplets may be born. Their mothers for protection against predators often hide the newborns in vegetation. They are weaned at around two to three months of age but may remain with their mothers for up to a year.

Juvenile Stage: During the juvenile stage, deer experience rapid growth and development. They learn to forage and navigate their environment, and young males may begin to grow antlers. This stage is crucial for the survival and health of deer, setting the foundation for their adult life.

Adult Stage: Deer reach sexual maturity at around one to two years of age, although this can vary between species and individuals. Adult deer have established home ranges and territories, with males often engaging in competitive behaviours for mating opportunities.

Senescence: As deer age, their physical condition declines. They may lose teeth, making it difficult to forage effectively, and males may experience a reduction in antler size. The ability to reproduce may also decline with age.

REPRODUCTIVE STRATEGIES

Mating Season (Rut): The mating season, or rut, typically occurs in the autumn, although the timing can vary between species and regions. During the rut, males engage in aggressive behaviours and competitions to secure mating opportunities with females.

Gestation and Birth: Following successful mating, the gestation period ranges from around 200 to 230 days, depending on the species. As the time of birth approaches, females seek secluded areas that provide cover and protection for the new-borns.

INFLUENCE ON POPULATION DYNAMICS

Reproductive Rate: The reproductive rate of deer populations significantly influences population dynamics. Factors such as age of sexual maturity, twinning rates, and the frequency of successful births contribute to the growth rate of deer populations.

Mortality Factors: Mortality factors such as predation, disease, starvation, and human-induced factors like road collisions and hunting play a crucial role in determining population dynamics. High juvenile mortality, for example, can significantly impact population growth rates.

Carrying Capacity: The carrying capacity of the habitat, which is the maximum population size that the environment can sustain indefinitely, is a crucial factor influencing deer population dynamics. Overpopulation can lead to habitat degradation, which in turn can affect the reproductive success and survival rates of deer.

Understanding the lifecycle and reproductive strategies of deer is instrumental in deer management. It aids in predicting population trends, assessing the impact of management interventions, and making informed decisions to ensure the sustainability of deer populations within their respective habitats. Through an informed approach, managers and stakeholders can work towards maintaining a harmonious balance between deer populations, their habitats, and the broader ecological community.

HABITAT PREFERENCES AND FOOD SOURCES

The ecological dynamics of deer within their habitats are shaped significantly by their preferences for certain environments and their dietary needs. These factors not only influence the daily and seasonal behaviours of deer but also have broader implications on the ecosystems they inhabit and their interactions with human communities.

HABITAT PREFERENCES

Woodland and Forests: Deer find solace in the dense canopies of woodlands and forests, which offer them protection from predators and adverse weather conditions. The understory vegetation provides a rich source of food, including a variety of shrubs, herbs, and tree saplings. The dense cover also offers secluded spots for birthing and rearing fawns, making these habitats crucial for the survival and propagation of deer populations. Woodlands with a mix of tree species and age classes provide a diverse habitat that meets the different needs of deer throughout the year.

Grasslands and Open Fields: Grasslands and open fields provide a contrasting habitat. The open terrain offers an abundance of grasses and forbs, which form a significant portion of a deer's diet, especially during the warmer months when these plants are in their growing phase. The openness also allows for better visibility, enabling deer to detect potential threats from predators. However, the lack of cover means deer may only frequent these areas during the day or when predator activity is low.

Wetlands and Marshes: The moisture-rich environment of wetlands and marshes offers a unique habitat. Besides providing water for drinking, these habitats host a variety of vegetation types, including aquatic plants, sedges, and rushes, which deer feed on. The dense reed beds also provide cover, making wetlands a favourable environment, especially for species like the Chinese Water Deer.

Heathlands and Moorlands: The rugged and often sparse vegetation of heathlands and moorlands provides a less favourable habitat. Nonetheless, deer can be found in these areas, particularly when other more suitable habitats are saturated or disturbed. The vegetation in these habitats, including heathers and gorse, provide an alternative food source, especially during winter when other food sources are scarce.

FOOD SOURCES

Grasses and Herbs: Grasses and herbs constitute a vital part of a deer's diet, providing essential nutrients and energy. During the growing season, deer extensively forage on these plants, gaining the necessary nutrition to support growth, reproduction, and the accumulation of fat reserves for winter.

Browse: Browsing on woody vegetation, including shrubs, bushes, and tree shoots, is particularly important during winter when other food sources are scarce. Deer have a preference for certain species, and their browsing activity can significantly impact vegetation composition and structure over time.

Fruits and Nuts: Seasonal availability of fruits and nuts offers a high-energy food source. Acorns, chestnuts, and berries are particularly favoured and provide essential fats and carbohydrates, aiding deer in building energy reserves.

Agricultural Crops: The abundance of food in agricultural fields can attract deer, leading to potential conflicts with farmers. Crops like soybeans, corn, and various vegetables provide nutritious forage, but deer feeding on these crops can cause significant economic losses to farmers.

BEHAVIOUR AND MOVEMENTS DRIVEN BY HABITAT AND FOOD

Seasonal Movements: The search for food and suitable habitat drives deer to exhibit seasonal migratory behaviours. Spring and summer bring them to higher elevations where vegetation is lush, while autumn and winter see them returning to lower elevations where snow cover is less and forage is more accessible.

Daily Movements: Deer are crepuscular animals, with their activity peaking during dawn and dusk. These daily movements are often dictated by foraging needs, predator activity, and environmental conditions such as temperature and weather.

Impact on Vegetation: The foraging behaviour of deer has a profound impact on vegetation. Over-browsing can lead to habitat degradation, altering vegetation composition, and structure, which in turn affect other wildlife species and the broader ecosystem.

Conflict with Humans: As deer seek out food, especially during harsh winter months, they may venture into agricultural fields and residential areas, leading to conflicts with humans. This behaviour underscores the importance of understanding deer habitat preferences and food sources in devising effective deer management strategies.

In essence, the habitat preferences and food sources of deer play a pivotal role in shaping their behaviours, health, and interactions with the environment and human communities. A thorough understanding of these aspects is fundamental to developing and implementing effective deer management strategies that aim to harmonise the co-existence between deer, other wildlife, and humans.

CHAPTER THREE

POPULATION MONITORING AND ASSESSMENT

POPULATION ESTIMATION TECHNIQUES

Understanding the population dynamics of deer is foundational to effective management strategies. Accurate population estimates are essential for monitoring population trends, assessing the impact of management actions, and ensuring the sustainability of deer populations. Various population estimation techniques have been developed, each with its own set of advantages and limitations. Here we delve deeper into these techniques, exploring their intricacies and their application in deer management.

Aerial Surveys: Aerial surveys entail counting deer from a vantage point above the ground, typically using helicopters, fixed-wing aircraft, or drones equipped with high-resolution cameras and sometimes thermal imaging technology.

Advantages:
Extensive Coverage: Aerial surveys are capable of covering vast expanses of land swiftly, providing a broader perspective on deer distribution and abundance.
Reduced Observer Bias: Utilizing standardized protocols and modern imaging technologies can minimize observer bias.

Limitations:
Cost: The deployment of aircraft and specialized equipment can be financially prohibitive.
Visibility Issues: Dense forest canopy or adverse weather conditions can obscure visibility, potentially leading to underestimations.

Ground Surveys: Ground surveys encompass a variety of techniques, including driving or walking transects, where observers count deer along predefined routes.

Advantages:
Cost-Effectiveness: Generally more affordable compared to aerial surveys.
Detailed Observations: Allow for closer inspection, possibly yielding more precise data on age, sex, and health of deer.

Limitations:
Time Consuming: Covering extensive areas can be labour-intensive and time-consuming.
Accessibility: Challenging terrain or dense vegetation can hinder accessibility and accuracy.

Camera Trapping: Strategically placed motion-activated cameras capture images or videos of deer, which can later be analysed to estimate population size.

Advantages:
Non-invasive Monitoring: Provides a means to monitor deer populations over time without human disturbance.
Behavioural Insights: Captures natural behaviours, aiding in understanding deer ecology.

Limitations:
Individual Identification: Distinguishing between individual deer can be challenging, affecting accuracy.

Equipment and Maintenance: Initial setup, maintenance, and data analysis can be time-consuming and costly.

Spotlight Surveys: Conducted during night time hours, spotlight surveys entail using powerful lights to illuminate and count deer in selected areas.

Advantages:
Increased Visibility: Deer are often more visible during night time, especially in open areas.
Cost-Effectiveness: Requires less specialized equipment compared to aerial surveys.

Limitations:
Disturbance: The presence of humans and vehicles might alter deer behaviour, potentially affecting count accuracy.
Limited Coverage: Coverage is limited to accessible areas and may not be representative of broader populations.

Pellet Group Counts: Counting pellet groups within defined plots to estimate deer density.

Advantages:
Cost-Effective: Requires minimal equipment and personnel.
Less Disturbance: Does not require direct observation or interaction with deer.

Limitations:
Decay Rates: Variability in faecal decay rates can affect accuracy.
Species Identification: Distinguishing pellet groups of different deer species can be challenging.

Track Counts: Analysing tracks in snow, mud, or sand to estimate deer activity and population size.

Advantages:

Low-Cost: Requires little to no specialized equipment.
Indirect Observation: Like pellet group counts, allows for population estimation without direct deer observation.

Limitations:
Environmental Dependency: Requires specific environmental conditions for success.
Identification and Count Accuracy: Accurate identification and count of tracks require expertise and favourable conditions.

Mark-Recapture Studies: Involves marking a subset of the population, then recapturing individuals over time to estimate population size using statistical models.

Advantages:
Accuracy: When conducted properly, can provide highly accurate estimates.
Movement and Survival Data: Provides valuable data on deer movement and survival rates.

Limitations:
Invasiveness: Capturing and handling deer can cause stress and potential injury.
Cost and Time Intensive: Requires significant resources for capturing, marking, and subsequent monitoring.

Each of these techniques offers unique insights into deer population dynamics, catering to different management objectives, geographical terrains, and budgetary constraints. The choice of method or a combination of methods hinges on the specific goals of the deer management program, the inherent advantages and limitations of each technique, and the available resources. Through adept application of these population estimation techniques, wildlife managers can garner the critical data requisite for informed, effective deer management.

HEALTH MONITORING

The health of deer populations is a pivotal concern in deer management, as it profoundly impacts the well-being of individual animals, population dynamics, and the broader ecosystem. Monitoring the health of deer involves a variety of techniques aimed at detecting diseases, assessing nutritional status, and identifying other health-related issues.

The data extracted from health monitoring activities aid in the timely detection and management of health problems, contributing to the overall efficacy of deer management programs. Below, we explore some of the prevalent methods used in health monitoring and discuss their significance in maintaining a robust deer population.

Physical Examinations: Conducting physical examinations of captured or harvested deer provide valuable insights into their health status. Advantages of this method include a direct assessment that enables a thorough evaluation of the deer's physical condition, including the assessment of body condition, dentistry, injury, and signs of disease. It also allows for immediate medical intervention if necessary (usual in park deer), such as treatment for parasites or injuries. However, it is invasive, as it requires capturing and handling deer, which can be stressful for the animals. Also, only a small proportion of the population can be examined, which may not be representative.

Necropsies: Performing necropsies on deceased deer to determine the cause of death and to identify any underlying health issues is another method. The advantages are that it provides comprehensive information on diseases and other health issues affecting deer populations and facilitates the identification of pathogens and other harmful agents. The limitations include it being a reactive method that only provides information after death has occurred, and it may over-represent individuals with diseases or other health problems due to sampling bias.

Parasite Monitoring: Evaluating the presence and load of external and internal parasites, through faecal examinations, skin scrapings, or blood tests, is crucial. Early detection enables the detection of parasite infestations, allowing for timely intervention, and some methods, like faecal examinations, are less invasive. However, specialized knowledge is required to accurately identify parasites and interpret results. Additionally, sensitivity may vary, potentially missing low-level infestations.

Disease Surveillance: Monitoring populations for signs of disease, including both active surveillance through regular health checks and passive surveillance through reports of sick or deceased animals, is essential. It facilitates early detection of disease outbreaks, enabling prompt management actions, and provides an ongoing assessment of population health, aiding in the identification of emerging health issues. However, it can be labour and resource-intensive, especially active surveillance programs. Passive surveillance relies on reporting, which can be biased or incomplete.

Nutritional Assessments: Assessing the nutritional status of deer through analysis of body condition, blood samples, and forage quality is also vital. It enables a thorough assessment of nutritional status, identifying deficiencies or imbalances, and provides insights into habitat quality and its impact on deer health. Yet, some methods, like blood sampling, are invasive and require capturing deer. Additionally, expertise and resources are required for sample collection and analysis.

Health monitoring is an indispensable component of deer management, ensuring that populations remain healthy and viable. It not only aids in the early detection and management of health issues but also provides a wealth of data that can be used to inform management decisions, improve habitat quality, and enhance the overall effectiveness of deer management programs. By investing in comprehensive health monitoring, wildlife managers can take a proactive approach to safeguard the health of deer populations, thus contributing significantly to the ecological integrity and sustainability of the habitats they occupy.

IMPACT ASSESSMENT ON FLORA AND FAUNA

The intricate interplay between deer populations and the ecosystems they inhabit necessitates a comprehensive assessment of their impacts on flora and fauna. The feeding habits, movement, and behaviours of deer can significantly shape the ecological landscapes they occupy. Understanding these impacts is fundamental to developing sustainable deer management strategies that harmonize the ecological, economic, and societal dimensions of wildlife conservation. Here, we explore into the various dimensions of deer impact on local ecosystems.

FOUR PRIMARY TYPES OF DEER DAMAGE

Browsing Damage: Browsing occurs when deer feed on the shoots of young trees or growing shrubs, stunting the growth of these plants in the process. From a fiscal standpoint, browsing is often considered the most detrimental as it tends to be widespread and can persist unnoticed for a substantial period. This stealthy damage can significantly affect profitability, often reaching severe levels before it catches the attention of the land manager. Browsing is a universal behaviour among deer, with both young and adult deer, regardless of gender, capable of inflicting this damage.

Fraying Damage: Fraying is a sporadic form of damage caused by male deer rubbing their antlers against the stem or trunk of trees, removing the bark in the process. Two notable peaks in fraying damage occur annually: during the shedding of velvet from fully grown antlers and, in some species, the marking of territories; and during the rut (mating season), where male deer, vying for and defending territories, exhibit aggressive behaviours through fraying and thrashing the branches and stems of growing shrubs.

Bole Scoring: Bole scoring, a behaviour primarily exhibited by Sika stags and to a lesser extent by Red deer stags, intensifies during the rut and the lead-up period. It involves stags pushing their brow tines into the bark of trees and lifting upwards, peeling the bark away. This behaviour aligns with territorial marking within the 'home range' or rutting areas occupied by Sika stags, signifying a form of territorial dominance.

Bark Stripping: Bark stripping is most commonly associated with Red deer, although Sika and Fallow deer may also exhibit this behaviour under certain conditions. The exact causes of bark stripping remain elusive, but it's often theorized to be a response to mineral deficiency. Instances of severe bark stripping have led to the total destruction of small plantations, particularly in enclosed herds with access to plantations.

Impact on Vegetation: Deer, by virtue of their herbivorous nature, are intricately linked to vegetation. Their browsing behaviour can lead to a plethora of changes within the vegetation dynamics of an ecosystem. Over-browsing, a common phenomenon in areas with high deer density, can lead to the depletion or over-exploitation of certain preferred plant species. This in turn affects the species composition and structural diversity of the vegetation, subsequently influencing the regeneration capacity of forests and other vegetated habitats. Additionally, the removal or reduction of vegetation cover can exacerbate soil erosion and alter hydrological processes, which can have far-reaching implications on the overall health, resilience, and functionality of the ecosystem.

Impact on Understory Vegetation: The understory vegetation often bears the brunt of deer browsing. Over-browsing can lead to a stark reduction in the diversity and abundance of understory plants. This dwindling of understory vegetation translates to a loss of crucial food and cover resources for a myriad of other wildlife species. The suppression of understory vegetation can hinder the recruitment and growth of tree and shrub species, potentially altering the successional trajectory of the ecosystem and affecting its long-term stability.

Impact on Tree Recruitment: Tree recruitment is a critical process in maintaining the sustainability and vitality of forest ecosystems. Deer browsing can significantly impact tree recruitment by feeding on seedlings and saplings. Over time, this behaviour can result in skewed forest age distributions, with fewer young trees emerging to replace the older, senescing ones. Additionally, the selective browsing by deer can foster a shift in tree species composition, favouring those species that are less palatable to deer while marginalizing others.

Impact on Other Wildlife: The reverberations of deer impact on vegetation extend to other wildlife species. The alteration in habitat structure and the availability of food resources can affect a wide spectrum of organisms, ranging from insects to large mammals. For example, the decline in understory vegetation can detrimentally affect ground-nesting birds, small mammals, and the insects that are dependent on specific plant species. Conversely, the creation of open areas through browsing can benefit species that thrive in open habitats or early successional vegetation stages.

Impact on Competitors: Deer populations can also influence the competitive interactions in an ecosystem. Deer share their habitat with other herbivores, and competition for food can ensue, especially during periods of food scarcity. In the UK deer do not currently have natural predators.

Interactions with Invasive Species: Deer browsing behaviour can inadvertently facilitate the invasion and proliferation of non-native plant species. By preferentially feeding on native plants, deer can provide a competitive advantage to invasive species, which may be less palatable or more resilient to browsing. This alteration in plant community composition can have far-reaching, often adverse, implications for the ecosystem, affecting native biodiversity and ecosystem functioning.

Human-Deer Conflicts: The interactions between deer populations and human activities often culminate in conflicts. For instance, deer browsing in agricultural fields can result in significant crop damage, leading to economic losses for farmers. Moreover, deer-vehicle collisions pose a serious threat to human safety and can result in substantial economic costs in terms of medical expenses and vehicle repairs.

A deer needs to eat about 6 to 8 percent of its body weight in green foliage and browse to stay healthy. For a 65kg deer, that's up to 4.5kg of food every day, which equates to 1.6 tonne per year.

The multi-dimensional impact of deer populations on local ecosystems underpins the necessity for a well-informed approach to deer management. By embracing a comprehensive impact assessment framework, wildlife managers and stakeholders can work in concert to develop and implement deer management strategies that balance the ecological needs of deer populations with the broader goals of biodiversity conservation, habitat restoration, and human-wildlife coexistence.

CHAPTER FOUR

HABITAT MANAGEMENT

HABITAT ENHANCEMENT TECHNIQUES

Optimizing habitat quality is pivotal for the conservation of deer and other wildlife, ensuring that ecosystems are equipped to support diverse and robust populations. A plethora of techniques are available for habitat enhancement, each designed to address specific challenges and improve particular aspects of the habitat. Here, we delve deeper into an array of habitat enhancement techniques, elucidating their methodologies and the potential benefits they confer to deer and other wildlife.

Forest Thinning and Prescribed Burning:

Detail: Forest thinning involves selectively removing trees to reduce canopy cover, enhance sunlight penetration, and stimulate the growth of understory vegetation. Prescribed burning is a controlled application of fire to eliminate accumulated leaf litter, control invasive species, and promote the regeneration of native plants.

Benefits: These techniques contribute to habitat diversification, providing varied food resources and shelter for deer and other wildlife. They also aid in controlling forest pests and diseases, fostering a healthier ecosystem.

Food Plots:

Detail: Food plots are cultivated areas planted with specific forage species that provide nutritional sustenance to deer and other wildlife. These plots can be planted with a mix of annual and perennial plants to ensure year-round availability of forage.

Benefits: Food plots augment the availability of nutritious food, particularly during lean periods, supporting healthier and more resilient deer populations. They also attract a variety of wildlife, promoting ecological interactions.

Native Vegetation Restoration:

Detail: This involves the reintroduction and cultivation of native plant species that have been displaced by invasive species or other disturbances. Native vegetation restoration aims at re-establishing the original plant community and enhancing habitat structure.

Benefits: Restored native vegetation offers an array of food and cover resources for deer and other wildlife, supporting biodiversity and improving ecosystem resilience.

Water Source Development:

Detail: Water source development entails creating or restoring water bodies like ponds, springs, and streams to ensure a reliable water supply for wildlife.

Benefits: Access to clean water is crucial for the survival and well-being of deer and other wildlife, especially during dry seasons. Water sources also create focal points of biodiversity within the landscape.

Riparian Area Protection and Restoration:

Detail: Riparian zones are transitional areas between terrestrial and aquatic ecosystems. Protecting and restoring these areas involve planting native vegetation, controlling erosion, and enhancing water quality.

Benefits: Riparian areas are biodiversity hotspots that provide a plethora of resources for deer and a myriad of other species, offering food, water, and shelter.

Wetland Restoration:

Detail: Wetland restoration aims at re-establishing the natural hydrology and vegetation of wetlands that have been drained or otherwise degraded.

Benefits: Restored wetlands provide habitat for a diverse array of species, including waterfowl, amphibians, and numerous invertebrates, enhancing the overall biodiversity of the area.

Brush Piling and Creating Cover:

Detail: Brush piling involves stacking cut branches and other vegetation to create dense, sheltered areas. Creating cover can also be achieved by planting dense shrubs and other vegetation.

Benefits: These structures provide essential cover for deer and other wildlife, offering refuge from predators and adverse weather conditions, and creating nesting and foraging sites.

Controlling Invasive Species:

Detail: Control measures such as mechanical removal, herbicide application, or biological control agents are employed to manage invasive plant species that degrade habitat quality.

Benefits: Controlling invasive species facilitates the recovery of native vegetation, improving the availability of natural food sources and habitat conditions for deer and other wildlife.

Erosion Control:

Detail: Techniques such as re-vegetation, installation of erosion control structures like silt fences, and improving soil quality are employed to control erosion and stabilize soil.

Benefits: Erosion control enhances soil quality and water retention, fostering a more hospitable environment for plant growth and, consequently, better habitat conditions.

Habitat Connectivity:

Detail: Habitat connectivity involves creating corridors or removing barriers to facilitate the movement and dispersal of deer and other wildlife between habitat patches.

Benefits: Enhanced connectivity allows for genetic exchange, access to seasonal habitats, and the colonization of new areas, promoting healthier and more resilient wildlife populations.

These habitat enhancement techniques are multifaceted, addressing various aspects of habitat quality and ecosystem health. By implementing a combination of these techniques, wildlife managers can work towards cultivating habitats that not only support thriving deer populations but also contribute to the broader conservation goals of biodiversity preservation and ecosystem resilience.

PREVENTING AND MANAGING HABITAT DEGRADATION

Addressing habitat degradation, a challenge often exacerbated by the overpopulation of deer and other wildlife, is a pivotal step towards ensuring the sustainability and biodiversity of ecosystems. The over-exploitation of resources, food and space, could lead to a discernible decline in habitat quality, adversely impacting not only the deer populations but also a myriad of other species sharing the same habitat.

Population Control: A cornerstone for preventing habitat degradation rests on the management of population sizes to ensure alignment with the carrying capacity of the habitat. Implementing methods such as controlled deer culls are pivotal in this regard. For instance, culling operations conducted in specific regions where deer populations have skyrocketed can help in reducing numbers to a sustainable level, thus averting over-browsing, habitat destruction, and the subsequent loss of biodiversity

Habitat Restoration: The essence of habitat restoration lies in rejuvenating degraded areas, a task achieved through the re-introduction of native vegetation, controlling erosion, and the improvement of soil and water quality. For instance, in areas overwhelmed by invasive species, reintroducing native vegetation such as deciduous trees and indigenous shrubs can significantly bolster the habitat's original fauna. Similarly, employing erosion control measures like the use of silt fences and gabions can prevent soil loss and improve water retention, thus enhancing the habitat's capacity to support wildlife populations. These restoration endeavours not only mitigate the impacts of overpopulation but also contribute to biodiversity and ecosystem services enhancement.

Monitoring and Adaptive Management: The strategy of continuous monitoring coupled with adaptive management practices is instrumental in identifying and addressing habitat degradation issues promptly. Employing modern technology like drones and camera traps can facilitate accurate monitoring of both wildlife populations and habitat conditions. This data, when analysed, provides the insight necessary to adapt management strategies to better address emerging challenges, thereby ensuring the long-term sustainability of wildlife populations and their habitats.

Fencing and Area Closures: In scenarios where immediate intervention is required to prevent further degradation, implementing fencing or area closures can be highly effective, but it does not solve the larger issue of overpopulation. For instance, fencing off riparian zones to prevent access by deer and other wildlife can allow vegetation to recover, thereby improving soil stability and water quality in these sensitive areas. These measures act as a respite for the habitat, allowing for vegetation recovery and the protection of soil and water resources, thus aiding in habitat restoration.

Public Education and Engagement: The role of public education and engagement is instrumental in fostering a culture of stewardship and responsible behaviour towards wildlife and their habitats. Through educational programs, field days, and informational campaigns, the public can be made aware of the impacts of overpopulation and the importance of habitat conservation. Such engagements can significantly enhance the effectiveness of conservation efforts and foster a sense of shared responsibility for preserving natural habitats.

Legal Framework and Enforcement: The establishment and enforcement of a robust legal framework are central to promoting sustainable wildlife management and habitat conservation. Legal measures such as laws regulating hunting, land-use changes, and the introduction of invasive species provide a structured approach to conservation. These measures, when enforced diligently, ensure compliance with best practices and standards aimed at preserving habitat integrity, thus contributing significantly to preventing and managing habitat degradation.

Alternative Food Sources: The provision of alternative food sources through food plots or supplemental feeding is a viable strategy to alleviate pressure on natural vegetation. For instance, food plots planted with a mix of cereals, legumes, and other forage crops can provide nutritious sustenance for deer, especially during periods of food scarcity. This mitigation of over-browsing pressure contributes to the prevention of habitat degradation.

Research and Innovation: Investing in research to develop innovative solutions and to better understand the dynamics between wildlife populations and habitat interactions is pivotal for informed decision-making. Technological advancements such as the use of drones for habitat monitoring and population surveys are examples of how innovation can significantly enhance habitat management efforts.

Each of these strategies contributes to a well-structured approach aimed at preventing and managing habitat degradation caused by overpopulation.

INTEGRATING DEER MANAGEMENT WITH OTHER LAND USES

The nexus between deer management and other predominant land uses such as agriculture, forestry, and urban development is intricately woven, demanding a nuanced, strategic approach to ensure a harmonious co-existence. Each land use, with its distinct objectives and operational paradigms, intersects with deer management in unique ways. Here we seek to delve deeper into the integration of deer management with these land uses, exploring strategies to foster mutual sustainability and reduced conflict.

Agriculture: Agricultural landscapes often become arenas of interaction between deer and human activities. The proximity of agricultural lands and deer habitats can lead to both beneficial and detrimental outcomes. Deer, finding sustenance in crop fields, might cause significant crop damage, translating into economic losses for farmers. The strategy to integrate deer management with agriculture hinges on minimising this damage while catering to the nutritional needs of deer. For instance, delineation of specific zones where deer can forage, away from critical crop areas, could be a viable solution. Additionally, implementing deterrents such as fencing or deploying scare devices **could** mitigate crop depredation. Establishing collaborative frameworks involving farmers, wildlife managers, and local communities could lead to mutually beneficial solutions. For instance, controlled hunting or culling initiatives on agricultural lands, conducted in a regulated manner, could help maintain deer populations at sustainable levels, thereby reducing crop damage.

Forestry: Forests are quintessential habitats for deer, providing shelter, food, and breeding grounds. However, the presence of deer can pose challenges to forestry operations, particularly in the realm of forest regeneration. Deer browsing on young saplings can significantly impede the natural regeneration process. A balanced approach to integrating deer management with forestry might encompass protective measures for young trees such as the use of tree shelters or fencing, alongside sustainable deer population management. A well-orchestrated logging plan that takes into account the habitat requirements of deer, ensuring a mix of age classes within the forest, could foster a symbiotic relationship between deer management and forestry operations.

Urban Development: The urban sprawl encroaching into traditional deer habitats presents a gamut of challenges and opportunities. Urban areas, with their relative lack of predators and hunting, can become unintended refuges for deer. However, this proximity can lead to human-deer conflicts, such as vehicle collisions, garden depredation, and potential transmission of diseases. A multifaceted approach is requisite for integrating deer management with urban development. This may span educating urban residents on co-existing with deer, implementing measures to mitigate deer-vehicle collisions. Thoughtful (if there is such a thing) urban planning that incorporates wildlife-friendly features such as green spaces, wildlife corridors, and deer-resistant landscaping can significantly ameliorate potential conflicts.

The venture of integrating deer management with other predominant land uses is a complex yet indispensable endeavour to ensure the sustainable co-existence of wildlife with agricultural, forestry, and urban landscapes. A collaborative, adaptive approach encompassing stakeholders from the wildlife management sector, agricultural and forestry industries, urban planning authorities, and local communities is essential to navigate the challenges and capitalize on the opportunities inherent in this integration. By fostering a shared vision and operational synergy among these diverse land use sectors, it's plausible to orchestrate a land use framework that is both sustainable and harmonious, ensuring the thriving co-existence of deer populations alongside agricultural, forestry, and urban developments.

CHAPTER FIVE

DEER POPULATION CONTROL STRATEGIES

CULLING AND SELECTIVE REMOVAL

The management of deer populations, particularly in areas where their numbers have burgeoned beyond sustainable levels, often necessitates the adoption of direct control measures. Among the repertoire of available strategies, culling and selective removal are two poignant approaches. These methods, while effective, require a meticulous evaluation of both ethical and practical considerations to ensure they are executed with the utmost regard for animal welfare, ecological integrity, and societal acceptance.

Culling: Culling, the reduction of animal populations through selective killing is a pragmatic approach to mitigate the impacts of overpopulation on ecosystems and human livelihoods. However, it is a venture laden with ethical considerations. The act of dispatching animals, even for management purposes, often elicits varied emotions and ethical dilemmas. It is imperative that culling operations are conducted with a clear, science-based rationale, ensuring that the benefits outweigh the ethical costs. The method of culling should adhere to the highest standards of animal welfare, minimizing suffering and ensuring a swift, humane dispatch.

The practical considerations surrounding culling are equally complex. The effectiveness of culling as a population control measure hinges on a comprehensive understanding of deer biology, behaviour, and the ecosystem dynamics. A culling operation requires skilled personnel, capable of making accurate judgments in the field to ensure the humane and effective removal of animals. Additionally, the cost, both financial and social, alongside the logistics and legal frameworks governing culling operations, require meticulous planning and community engagement.

Selective Removal: Selective removal transcends mere population reduction; it entails the targeted removal of specific individuals or groups within a population based on defined criteria such as age, sex, or health status. This approach aims at modifying the population structure to achieve specific management objectives, like reducing reproductive potential or removing individuals prone to causing conflicts.

The ethical considerations surrounding selective removal are akin to those of culling, albeit with an added layer of complexity due to the targeted nature of removal. The criteria for selection must be grounded in sound scientific evidence to justify the removal of particular individuals or groups. The welfare of the animals and the potential impact on social structures within the deer population warrant careful consideration.

Practically, selective removal demands a high level of expertise to accurately identify and remove the targeted individuals. Monitoring the impacts of selective removal on the deer population and the broader ecosystem is paramount to evaluate the effectiveness of the strategy and make necessary adjustments.

In essence, both culling and selective removal are potent tools in the arsenal of deer management strategies. However, their implementation demands a rigorous ethical examination, robust practical planning, and an overarching framework of transparency and community engagement. Through a blend of ethical consideration, scientific evidence, and practical feasibility, culling and selective removal can be employed to foster the sustainable management of deer populations, ensuring the co-existence of deer with other land uses and the natural environment.

EXCLUSION AND DETERRENTS

Among the various strategies, exclusion and deterrent measures emerge as proactive approaches to modulate deer movement and behaviour while reducing potential conflicts. Here we will look at the various facets of physical barriers and deterrents, evaluating their efficacy and outlining considerations for their implementation.

PHYSICAL BARRIERS

Physical barriers serve as a tangible means to deter deer from entering specific areas, thereby minimizing interactions that could lead to conflicts or economic losses. Fencing, the cornerstone of physical exclusion measures can be designed in a multitude of ways tailored to the specific challenges posed by deer.

Electric Fencing: Electric fences deter deer through the delivery of an aversive electric shock, making them a potent deterrent. The design can range from simple single-strand electric fences to multi-strand or mesh designs. However, the maintenance of electric fences, including ensuring a consistent power supply and the repair of breaches, is a pertinent consideration. The initial installation costs, lack of effective data and the ongoing power requirements add to the operational expenses and choice.

Solid Fencing: Solid fences, constructed from wood, metal, or other durable materials, provide both a visual and physical barrier to deer. Their efficacy hinges on their height and robustness, as deer possess a remarkable jumping ability. A height of at least eight feet is often recommended to effectively deter deer. The aesthetic impact, durability, and cost are significant considerations when opting for solid fencing.

Temporary Fencing: In certain scenarios, temporary fencing might suffice, especially during critical periods like the growing season in agricultural areas. Temporary fencing can be more cost-effective and less labour-intensive to install, yet may offer a lower level of deterrence compared to permanent fencing structures.

DETERRENTS

Deterrents aim to exploit the sensory perceptions of deer to evoke aversive responses, encouraging them to avoid specific areas. The selection of deterrents span across olfactory, auditory, and visual domains, each with its own set of advantages and challenges.

Olfactory Deterrents: By leveraging the acute sense of smell of deer, olfactory deterrents use repellents that mimic the scents of predators or other odorous substances deemed unpleasant by deer. However, the efficacy may wane as the scent dissipates, necessitating regular re-application. Some olfactory deterrents might also be unpleasant to humans or other non-target species.

Auditory Deterrents: Auditory deterrents employ noises, potentially of predators or other alarming sounds, to scare away deer. The surrounding noise levels and the deer's familiarity with various sounds might influence the effectiveness. Habituation is a significant challenge, as deer may become accustomed to the noises if not paired with a real threat.

Visual Deterrents: Visual deterrents utilize flashing lights, moving objects, or other visual cues to create an environment perceived as threatening by deer. The mobility of deer and their ability to quickly adapt to new stimuli might result in diminished effectiveness over time.

The deployment of exclusion and deterrent measures requires a deep understanding of the deer behaviour, the local context, and the broader implications on the community and other wildlife. The cost-effectiveness, maintenance requirements, and potential societal acceptance are pivotal considerations that will influence the practicality and success of these measures. Through an informed, adaptive, and collaborative approach, the implementation of physical barriers and deterrents can, when used correctly significantly contribute to the sustainable management of deer populations.

CHAPTER SIX

ETHICAL CONSIDERATIONS

ETHICAL CULLING PRACTICES

Engaging in a dialogue surrounding the ethical and practical dimensions of culling, especially in the realm of deer management, form the basis of such an intervention in wildlife populations. Here we look into the moral and practical dimensions of ethical culling, exploring humane techniques, and unearthing the broader considerations that come to bear upon this management strategy.

MORAL DIMENSIONS

The moral consideration of culling is a profound one; the linchpin of the moral argument rests on the provision of a robust, science-based rationale that delineates the necessity and the benefits of culling, either from an ecological or societal vantage point. This includes an understanding of the ecological ramifications of deer overpopulation, the impacts on human livelihoods and safety, and the anticipated ecological restoration post-culling intervention.

Societal acceptance of culling hinges on a number of factors including cultural predilections, educational background, and individual ethical standpoints. The dialogue extends into the legal and policy realms, underscoring the necessity for a coherent, transparent, and ethically grounded regulatory framework that governs wildlife management endeavours including culling.

PRACTICAL DIMENSIONS

The practical aspect of ethical culling is varied, encompassing a number of factors that are critical to the operational integrity and humaneness of the culling process.

Skilled Personnel: The aptitude of the individuals tasked with the culling operation is a cornerstone for ensuring humaneness and efficacy. Highly skilled and trained hunters, with a profound understanding of deer anatomy and behaviour, are pivotal to ensuring that the culling is conducted swiftly and humanely.

Equipment and Techniques: The choice of equipment and the techniques employed are instrumental in defining the humaneness of any culling activity. Equipment should be meticulously maintained, and the techniques should be honed to ensure a swift and humane demise of the animals. The calibre of firearms, the type of ammunition, and the accuracy of shooting are critical determinants that contribute to the humaneness and effectiveness of the culling operation.

Monitoring and Evaluation: A rigorous and continuous monitoring and evaluation framework is indispensable. This includes real-time monitoring of culling operations, post-mortem examinations, and evaluation against predefined humane indicators. Such a framework can yield invaluable insights into the effectiveness and humaneness of the culling techniques, fostering a culture of continuous improvement.

Transparency and Accountability: The ethos of transparency and accountability is cardinal in engendering societal acceptance and trust. Open engagement with the public and stakeholders, coupled with clear, factual communication about the necessity, procedures, and outcomes of culling operations, is instrumental in building a foundation of understanding and trust.

COMMUNITY ENGAGEMENT AND PUBLIC PERCEPTIONS

The narrative surrounding deer management is an expansive one, intertwining ecological, legal, and societal threads. The perceptions and engagement of the community and the broader public are pivotal, as they significantly influence the acceptance, support, and, consequently, the success of deer management initiatives. Here we look into the nuances of community engagement and public perceptions, their impact on deer management practices and the broader goals of fostering co-existence between human communities and deer populations.

Public Perceptions: The public perceptions towards deer and their management is complex, often influenced by a confluence of factors including cultural values, personal experiences, educational backgrounds, and the narrative presented through various media channels. Understanding the breadth and depth of public perceptions is not merely an exercise in social analysis but a critical step towards crafting deer management strategies that resonate with the societal values and concerns.

Community Engagement: The engagement of local communities and stakeholders forms the cornerstone of socially robust and effective deer management initiatives. The spectrum of engagement is vast, encompassing informational outreach, public consultations, collaborative decision-making processes, and community-based deer management programs.

1. **Informational Sessions:** Orchestrating informational sessions to disseminate accurate, comprehensive, and accessible information about deer ecology, the challenges

engendered by overpopulation, and the objectives and methodologies of management interventions is a foundational step in fostering public understanding and garnering support.

2. **Public Consultations:** Public consultations serve as a democratic platform for stakeholders and community members to articulate their views, concerns, and suggestions regarding deer management. This approach nurtures a sense of ownership, inclusivity, and trust in the decision-making processes.

3. **Collaborative   Decision-making:**   Collaborative decision-making processes, encompassing a diverse array of stakeholders including local communities, wildlife managers, policymakers, and other interested parties, can engender more robust, locally accepted, and effective deer management strategies. The synergy of varied perspectives can lead to a more nuanced and holistic approach to deer management.

4. **Community-Based Deer Management Programs:** Community-based deer management programs are predicated on the empowerment of local communities to take an active role in deer management. By leveraging local knowledge, resources, and support, these programs can foster a sense of stewardship, enhance the effectiveness and acceptance of deer management interventions, and foster a culture of co-responsibility.

For a good example visit: https://highweald.org/

Educational Programs: Educational programs are instrumental in enlightening the public about the intricacies of deer ecology, the rationale for management interventions, and the humane and ethical practices employed. Through education, misconceptions can be dispelled, fears allayed, and a culture of understanding, empathy, and respect for wildlife can be nurtured.

Media and Communication Strategies: Effective media and communication strategies play a crucial role in shaping public perceptions and fostering support for deer management initiatives. Transparent, factual, and timely communication through a multitude of media channels can build trust, dispel misinformation, and create a conducive environment for constructive dialogue and collaboration.

Feedback Mechanisms: Establishing robust feedback mechanisms to gauge public response, address concerns, and adapt management strategies based on public input is essential for ensuring the social legitimacy and success of deer management initiatives. These mechanisms provide a channel for continuous dialogue, adaptation, and improvement, enhancing the social resonance and efficacy of deer management practices.

The path towards effective, humane, and socially accepted deer management practices is significantly smoothened through a concerted effort aimed at engaging the community, understanding public perceptions, and fostering a culture of inclusivity and collaboration. This approach augments the effectiveness of deer management initiatives and resonates with the broader societal ethos of democracy, transparency, and co-existence with wildlife.

CHAPTER SEVEN

HEALTH AND DISEASE MANAGEMENT

COMMON DEER DISEASES

The vitality and health of deer populations are intricately linked with broader ecological and societal concerns. Among the varied factors impacting deer health, diseases are significant concerns that reverberate beyond the affected individuals to populations, other wildlife species, and at times, humans. Here we seek to shed light on diseases afflicting deer, their implications, and the diverse prevention strategies aimed at mitigating the risks and impacts associated with these diseases.

Chronic Wasting Disease (CWD): Chronic Wasting Disease is a formidable foe in the realm of deer management. This fatal neurodegenerative disorder, akin to Bovine Spongiform Encephalopathy (BSE) in cattle or Creutzfeldt-Jakob disease in humans, manifests as progressive weight loss, behavioural abnormalities, and eventually, death. The disease is insidious, with a prolonged incubation period, during which infected deer can spread the disease. The contagious nature and the lack of a known cure make CWD a significant threat to deer populations. *Thus far, there have been no reported cases of CWD or other TSE in deer in Great Britain (GB). This is based on surveys of wild*

*and farmed red deer (Cervus elaphus elaphus) (EFSA, 2011).
Given the consequences of CWD observed in North America, it
is of high importance that GB remains free of the disease.*

Bovine Tuberculosis (bTB): Bovine Tuberculosis, propelled by
the bacterium Mycobacterium bovis, is a chronic bacterial disease
that can afflict deer. The disease manifests with symptoms such
as coughing, weight loss, and the formation of granulomas in
the lungs and other tissues. The potential zoonotic implications
and risks to livestock highlight the importance of controlling
bTB in deer populations. Control measures encompass regular
monitoring, testing, and, in some scenarios, culling of infected
individuals to staunch the spread of bTB. Biosecurity measures,
such as restricting the movement of infected animals and proper
disposal of carcasses, are also pivotal in bTB control.

Foot and Mouth Disease (FMD): Foot and Mouth Disease,
driven by the Picornavirus, is a highly contagious viral ailment
afflicting cloven-hoofed animals such as cattle, swine, and
deer. It unveils itself through symptoms like fever, lameness,
and the emergence of vesicles or blisters in the mouth and on
the feet. The rapid transmission rate of FMD underscores the
critical necessity of stringent control and preventive measures to
safeguard livestock and wildlife populations. The containment
strategies encompass immediate reporting, quarantine of
infected and susceptible animals, and, in many instances, culling
to arrest the spread of the virus.

Epizootic Haemorrhagic Disease (EHD): Epizootic
Haemorrhagic Disease is a virulent ailment inflicting deer,
characterized by extensive haemorrhages, fever, and often
death. The disease is vectored by biting midges and often sees
outbreaks in late summer and fall. Preventive measures extend
into habitat management to curtail breeding sites for midges
and vaccination programs where available. The detrimental
impact on deer populations, especially during severe outbreaks,

underscores the importance of preventive and control measures against EHD.

Parasitic Infections: Deer are susceptible to a number of parasitic infections, including gastrointestinal worms, lungworms, and ectoparasites such as ticks and mites. These parasitic adversaries can precipitate poor body condition, reduced fertility, and increased susceptibility to other diseases.

Blue Tongue: Blue Tongue, another viral disease vectored by biting midges, is characterized by swelling of the head and neck, haemorrhages, and lameness. Like EHD, managing the habitat to control midge populations and vaccination are key preventive strategies against Blue Tongue.

PREVENTION STRATEGIES

The landscape of prevention strategies is broad and often intersects with broader deer management objectives.

1. **Monitoring and Surveillance:** Robust monitoring and surveillance programs is the linchpin for early detection of diseases, facilitating timely interventions to contain disease spread.

2. **Vaccination:** Vaccination programs, usually employed within fenced deer, where vaccines are available and applicable, can significantly truncate the incidence and impact of certain diseases.

3. **Habitat Management:** Habitat management strategies, such as controlling vectors through water management or vegetation control, can play a significant role in disease prevention.

4. **Public Education:** Educating the public, especially deer

managers and farmers, about deer diseases, their risks, and the measures to prevent or contain them, is crucial for a collaborative approach to disease management.

5. **Biosecurity Measures:** Stringent biosecurity measures such as restricting the movement of potentially infected individuals, using protective gear, and proper disposal of carcasses can significantly curb the risk of disease transmission.

Through a concerted effort aimed at understanding, preventing, and managing deer diseases, the pathway towards healthy deer populations and minimized risks to other wildlife, livestock, and humans can be significantly enhanced.

CHAPTER EIGHT

HUMAN-DEER CONFLICTS

TRAFFIC ACCIDENTS

The intertwining of deer habitats and human domains often manifests in the occurrence of traffic accidents involving deer, which not only pose significant safety risks to humans but also result in deer fatalities. The propensity of deer to venture onto roadways, especially during dawn and dusk, often culminates in accidents that can have severe or even fatal repercussions. Here we aim to unravel the prevalence of deer-related traffic accidents and the various preventive measures that can be deployed to mitigate these incidents, thereby fostering a safer coexistence between deer populations and human communities.

Prevalence: Deer-related traffic accidents predominantly surge during the breeding season, and when other habitat based disturbance is heightened. The expansion of urban areas into deer habitats and the proliferation of road networks through these habitats accentuate the chances of encounters between deer and vehicles. The occurrence of accidents is not a mere sporadic event but often exhibits a pattern, with certain areas known as accident hotspots due to recurrent incidents. The impact of these accidents is multifold, encompassing human injuries or fatalities, property damage, and deer mortalities. The financial burden associated with deer-related traffic accidents is also substantial, reflecting in vehicle repair costs, medical expenses, and insurance claims.

There are up to 75,000 Deer Vehicle Collisions each year in the UK resulting in 400 to 700 human injuries and several human fatalities each year - https://www.deeraware.com/

PREVENTION MEASURES

Preventing deer-related traffic accidents requires a collaborative and multi-faceted approach that combines engineering, education, and ecology.

Engineering Solutions: Engineering solutions form the cornerstone of preventive measures. These include the installation of deer fencing along roadways, especially in accident-prone areas, to deter deer from entering the roadway. Additionally, underpasses and overpasses can be constructed to provide safe passage for deer across busy roads. Well-designed and strategically placed wildlife crossing structures can significantly reduce deer-vehicle collisions.

Warning Systems: Implementing warning systems such as dynamic warning signs that alert drivers about deer presence, especially in high-risk areas, is crucial. These systems can be activated by sensors that detect deer movement near roadways, thereby providing real-time warnings to drivers.

Vegetation Management: Managing vegetation near roadways to reduce the attractiveness of these areas to deer and to improve visibility for drivers is instrumental in accident prevention. Clearing vegetation that serves as food or cover for deer near roadways can dissuade deer from venturing close to traffic zones.

Public Education: Educating the public about the risks of deer-related traffic accidents and providing guidance on how to react when deer are spotted on or near roadways is essential. Public awareness campaigns can significantly contribute to reducing accidents by promoting safe driving practices, especially during high-risk periods.

Road Planning and Design: Integrating wildlife considerations into road planning and design is pivotal. This could encompass avoiding the construction of new roads through crucial deer habitats or migration corridors, and designing roads to minimize habitat fragmentation and deer-vehicle interactions.

Monitoring and Research: Continual monitoring of deer-related traffic accidents and research to understand deer movement patterns and behaviour in relation to roadways can furnish valuable insights. These insights can guide the enhancement of preventive measures and the development of innovative solutions to reduce accidents.

Through the amalgamation of these preventive measures, the paradigm of reducing deer-related traffic accidents is progressively actualized. The multi-pronged approach not only minimizes the risks associated with deer-vehicle collisions but also underscores the imperative of fostering a harmonious coexistence between deer populations and human communities. This proactive approach towards preventing deer-related traffic accidents aligns with the broader ethos of sustainable deer management, veering towards a scenario where both deer populations and human communities can thrive amidst reduced conflict and enhanced safety.

AGRICULTURAL AND FORESTRY DAMAGE

The presence of deer in regions interspersed with agricultural lands and forests is a scenario ripe for conflicts. Deer, in their quest for sustenance, often venture into agricultural fields and forests, leading to notable damage. Their feeding habits can result in significant economic losses for farmers and forest managers. Here we delve into the impact of deer on agriculture and forestry and elucidates the spectrum of management strategies that can be employed to mitigate the damage, fostering a balanced coexistence.

Impact on Agriculture: Deer pose a considerable challenge to agricultural endeavours. Their foraging activities often result in the consumption of crops, leading to reduced yields. The economic repercussions can be substantial, with farmers bearing the brunt of the loss in crop productivity and the consequent financial strain. Additionally, the damage inflicted on crops can affect the quality of produce, further exacerbating the economic impact. Deer also have the potential to transmit diseases to livestock, presenting another dimension of conflict and economic loss.

Impact on Forestry: Forests too bear the mark of deer presence. Deer browsing and bark stripping activities can severely impact tree regeneration, growth, and survival. The damage to young saplings and understory vegetation can result in long-term adverse effects on forest structure, composition, and biodiversity. The alteration of habitat through deer browsing can have cascading effects on other wildlife species that share the same habitat, potentially disrupting the ecological equilibrium.

URBAN DEER ISSUES

The urban realm, with its unique blend of natural and man-made elements, presents a setting that often attracts deer populations. The lure of abundant food sources, coupled with the lack of natural predators, makes urban areas appealing havens for deer. However, the presence of deer in urban settings creates a number of challenges, not only for the deer themselves but also for the human inhabitants.

CHALLENGES

Human-Deer Conflicts: The interaction between humans and deer in urban settings often culminates in conflicts. Deer may venture into gardens, causing damage to vegetation, or traverse across roads, leading to traffic accidents. Furthermore, the potential for disease transmission from deer to humans or pets adds another layer of conflict.

Public Safety Concerns: Public safety concerns arise from deer-related traffic accidents and aggressive behaviour exhibited by deer, especially during the rut or when females are protecting their fawns. These situations can pose significant safety risks to humans.

Overpopulation: The absence of natural predators and the availability of food contribute to deer overpopulation in urban areas. Overpopulation exacerbates the aforementioned challenges and may lead to malnourishment and disease outbreaks within the deer population due to the exceeding of the area's carrying capacity.

SOLUTIONS

Community Engagement: Engaging the community in urban deer management programs fosters a collaborative ethos. Through education and involvement, community members can become integral participants in addressing urban deer challenges. Public awareness campaigns can illuminate the challenges and solutions associated with urban deer populations, thereby cultivating a more informed and cooperative populace.

Habitat Modification: Modifying urban habitats to deter deer presence can be effective. This may include the removal of attractants such as certain types of vegetation, and the use of deer-resistant plants in landscaping. Additionally, the management of green spaces to reduce cover and the availability of food sources for deer can contribute to making urban areas less appealing to them.

Population Control: Population control through culling can help in managing urban deer numbers. However, these measures require careful consideration of ethical, ecological, and logistical factors.

Physical Barriers: Employing physical barriers such as fencing or netting around gardens and other sensitive areas can deter deer access and mitigate damage. Designing roads with wildlife in mind, including the incorporation of wildlife crossings, can also mitigate the risks associated with deer-vehicle collisions.

By combining together a fabric of solutions encompassing community engagement, habitat modification, population control, and continuous learning, the urban deer conundrum can be navigated.

CHAPTER NINE

ECONOMIC ASPECTS OF DEER MANAGEMENT

COST-BENEFIT ANALYSIS

The economics of deer management is a complex narrative intertwining costs and benefits, each with direct and indirect reverberations on communities, ecosystems, and the deer populations themselves. An astute assessment of the economic dynamics is pivotal for informed decision-making and the rational allocation of resources in deer management endeavours. Here we look at the cost-benefit analysis approach, providing a lens through which the economic dimensions of different deer management strategies can be evaluated, fostering a more balanced understanding of the economic landscape.

Direct Costs: The direct costs associated with deer management are often conspicuous and quantifiable. They encompass the financial outlays required for the implementation of various management strategies such as culling, fencing, repellents, public education campaigns, and habitat restoration initiatives. These costs also include personnel expenses, equipment purchases, and the maintenance of infrastructure such as fencing and wildlife crossings.

Indirect Costs: Indirect costs, though less palpable, hold significant economic weight. They encapsulate the economic losses incurred due to deer induced agricultural and forestry damage, deer-vehicle collisions, and the potential transmission of diseases to livestock or humans. The administrative overheads, including the costs associated with policy formulation, enforcement, monitoring, and research, also fall within the ambit of indirect costs.

Benefits: On the flip side, deer management ushers in a suite of benefits. Effective management can lead to reduced agricultural and forestry damage, lower incidence of deer-vehicle collisions, and a diminution in disease transmission risks. The enhancement of habitat quality and the preservation of biodiversity are long-term benefits with broad ecological and societal implications. Additionally, deer management can foster recreational and tourism opportunities, including hunting and wildlife watching, which can inject economic vitality into local communities. The sustainable harvest of deer also opens avenues for venison production, contributing to local economies.

Cost-Benefit Evaluation: The crux of the cost-benefit analysis lies in the meticulous evaluation of the economic costs and benefits associated with each deer management strategy. This evaluation necessitates a thorough understanding of the direct and indirect costs, work alongside the array of short-term and long-term benefits. The aim is to discern the net economic impact and to navigate towards strategies that deliver maximum benefits with minimum costs.

Methodologies: Various methodologies can be employed in conducting a cost-benefit analysis. These methodologies could encompass the use of economic modeling, stakeholder consultations, and the analysis of historical and projected data. Furthermore, the application of sensitivity analysis can provide insights into how variations in key parameters affect the overall cost-benefit dynamics.

The cost-benefit analysis fosters a pragmatic approach towards deer management, steering towards strategies that are not only ecologically sound but also economically rational. By shedding light on the economic underpinnings, the analysis provides a robust framework for decision-making, ensuring that a clear understanding of the economic costs and benefits support the allocation of resources in deer management initiatives. This rational economic lens, amalgamated with ecological and societal considerations, propels the narrative of sustainable deer management forward, contributing to the broader discourse on wildlife management in a nuanced and informed manner.

UK FUNDING AND GRANTS

The sphere of deer management requires substantial financial inputs to address the challenges and implement effective management strategies. In the UK, a variety of funding opportunities and grants are available to support deer management initiatives. These financial provisions play a critical role in supporting the efforts of landowners, local authorities, and wildlife management agencies. Here we explore the array of funding opportunities and grants that could be harnessed to underpin deer management initiatives, fostering a financially sustainable approach to managing deer populations and mitigating their impact.

Please note these are constantly changing therefore you should reach out to the relevant department for updated details.

Government Grants: The UK government, through various departments and agencies, offers grants to support deer management activities. These grants may cover a wide spectrum of initiatives including habitat restoration, population control, research, and public education campaigns. The criteria for grant eligibility and the application process may vary across different government entities, thus necessitating a thorough understanding of the stipulations and requirements set forth by the granting institutions.

Non-Governmental Organisations (NGOs) and Trusts: Numerous non-governmental organisations and trusts also extend financial support for deer management. These entities often have a vested interest in wildlife conservation, habitat preservation, and community engagement. The grants provided by NGOs and trusts can be instrumental in facilitating deer management projects, especially those with a strong emphasis on ecological preservation and community involvement.

European Union Funding: Historically, the European Union has been a source of funding for wildlife management initiatives, including deer management, in the UK. Even with the UK's departure from the EU, there may still be opportunities for funding from European sources for transboundary or international cooperative wildlife management projects.

Industry Partnerships: Engaging with industry partners can also open channels for funding. Companies and corporations, particularly those with a stake in agriculture, forestry, or land development, may have interests in supporting deer management initiatives as part of their corporate social responsibility or to mitigate deer-related challenges affecting their operations.

Local Community Funding: Local community funding, often garnered through fundraising campaigns or local grants, can be a viable source of financial support. Community-driven funding initiatives underscore the collective resolve to address deer management challenges and often foster a sense of ownership and engagement among the local populace.

Research Institutions: Research institutions may also offer funding for deer management, especially for projects that contribute to the scientific understanding of deer ecology, behaviour, and management. Collaborating with research institutions can not only secure funding but also foster a data-driven approach to deer management.

The availability of funding and grants is a key factor for advancing deer management objectives. By leveraging these financial resources, stakeholders in deer management can better equip themselves to address the challenges posed by deer populations. This financial augmentation is fundamental in propelling deer management initiatives forward, thereby contributing to the broader goal of fostering a sustainable and balanced interaction between deer populations, human communities, and the natural environment.

VENISON PRODUCTION AND MARKETING

Please note these are constantly changing therefore you should reach out to the relevant department for updated details.

The production and marketing of venison in the UK are

underpinned by a comprehensive set of legal frameworks designed to assure the safety, quality, and ethical processing of deer products. These regulations are instrumental in maintaining public health standards, ensuring animal welfare, and promoting sustainable deer management practices.

Specific Hygiene Rules: The crux of the regulation surrounding the production of venison hinges on adherence to specific hygiene rules. These rules, enshrined in Retained Regulation (EC) No. 853/2004 for Great Britain, encapsulate a broad spectrum of hygiene requirements tailored for businesses engaged in the production of food of animal origin. They delineate the hygiene standards necessary at every stage of production, processing, and distribution of venison to prevent any risk to public health.

General Rules on Animal By-products: Beyond the primary products, the regulatory ambit also extends to animal by-products not destined for human consumption. The guidelines, as stipulated in Retained Regulation (EC) No. 1069/2009 for Great Britain, oversee the handling and processing of these by-products, ensuring they are managed in a manner that poses no threat to public health or the environment.

Hygiene Requirements: The generic hygiene requisites applicable to all food businesses are also pertinent to those dealing with venison. These mandates, encapsulated in Retained Regulation (EC) No. 852/2004 for Great Britain, provide a blueprint for hygiene practices across the food production chain, ensuring a baseline of hygiene measures that safeguard food quality and consumer health.

Official Controls and Activities: Ensuring adherence to the diverse food and feed laws, animal health and welfare regulations, plant health guidelines, and plant protection products mandates requires a robust system of official controls and activities. These controls, detailed in Retained Regulation (EU) No. 2017/625 for Great Britain, provide the framework for official inspections and other activities aimed at enforcing the application of food and feed law, thus ensuring a holistic compliance with the regulatory landscape.

General Principles of Food Law: The overarching umbrella of food law principles and requisites, including the pivotal aspect of food and feed traceability, is enshrined in Retained Regulation (EC) No. 178/2002 for Great Britain. These principles lay down the broad framework of food law, setting the stage for a structured and coherent regulation of food safety, quality, and ethical considerations throughout the food chain.

Venison Supply Legislation: The Deer (Scotland) Act 1996 and the Food Hygiene Regulations play a significant role in shaping the legal framework associated with the supply of venison. They elucidate the obligations and responsibilities incumbent upon individuals and entities involved in the venison supply chain, ensuring a well-regulated and transparent system of venison supply.

Meat Products Legislation: The regulatory landscape also encompasses specific legislation for products containing meat. The Food Information Regulations 2014 and Products Containing Meat etc. (England) Regulations 2014 provide guidelines on the types of meat products and the terms used in marketing such products, ensuring clarity and transparency in the marketing and labelling of meat products.

Quality Assurance and Food Safety: The UK's approach to deer management, food safety, and legal protections is geared towards ensuring humane culling of deer and rigorous quality assurance. Before venison is deemed fit for consumer consumption, it must undergo stringent handling and processing requirements. These mandates are pivotal in assuring the safety, quality, and ethical standards of venison products, thereby engendering consumer trust and promoting the sustainable management of deer populations.

The synergy of these regulatory frameworks ensures that venison production and marketing are conducted in a manner that upholds public health standards, assures animal welfare, and promotes ethical and sustainable practices in deer management.

CHAPTER TEN

COMMUNITY ENGAGEMENT AND EDUCATION

PUBLIC AWARENESS CAMPAIGNS

Public Awareness Campaigns are pivotal in creating a supportive and informed community for the successful implementation of deer management initiatives. These campaigns serve as a medium to educate the public about the ecological and societal importance of deer management, exploring its impact on biodiversity, habitats, and human communities.

Educational Value: Public awareness campaigns hold immense educational value as they disseminate accurate information about deer populations, their impact on the environment, and the necessity of management interventions. Through these campaigns, misconceptions can be corrected, and a factual understanding can be fostered among the public. The educational facet of these campaigns is instrumental in illuminating the challenges posed by overpopulation and the benefits of responsible deer management practices. By providing a clear, fact-based narrative, these campaigns contribute to a well-informed public discourse, which is crucial for garnering support for deer management initiatives.

Promoting Community Engagement: Community engagement is a significant aspect of successful deer management initiatives. Effective public awareness campaigns provide avenues for public participation, encouraging community members to take an active role in deer management efforts. These campaigns can galvanize community support, foster collaborative efforts, and create a sense of shared responsibility among different stakeholders. By promoting a culture of engagement and stewardship, public awareness campaigns contribute to the sustainability and effectiveness of deer management programs.

Shaping Public Perception: Public perception significantly influences the level of support for deer management initiatives. Through well-crafted awareness campaigns, the narrative surrounding deer management can be shaped to underscore the ethical, ecological, and societal imperatives of these initiatives. Highlighting the benefits of deer management, addressing potential concerns, and showcasing successful case studies can help foster a positive public perception. A favourable public perception is instrumental in garnering broader support and ensuring the acceptability of deer management initiatives.

Support for Policy and Funding: Public support, as galvanized through effective awareness campaigns, can translate into political support and funding for deer management initiatives. When the public is well-informed and supportive, it creates a conducive environment for advocating favourable policies, securing funding, and ensuring the continuity and expansion of deer management programs. The support generated through these campaigns can be a catalyst for policy change and increased funding, which are crucial for the long-term success of deer management initiatives.

Enhancing Compliance and Cooperation: Compliance with deer management regulations and cooperation among landowners, local communities, and deer management authorities are enhanced through effective public awareness campaigns. A well-informed public is more likely to comply with regulations, participate in monitoring and reporting activities, and cooperate with management efforts. By fostering a culture of compliance and cooperation, these campaigns contribute significantly to the effectiveness and efficiency of deer management initiatives.

Feedback Mechanism: Public awareness campaigns often serve as a two-way communication channel, enabling authorities to receive feedback from the community. This feedback is invaluable in assessing the effectiveness of deer management strategies, identifying areas of improvement, and adapting strategies to better meet the needs and expectations of the community. Through this feedback mechanism, deer management authorities can fine-tune their strategies, ensuring they are aligned with community needs and preferences.

The well-orchestrated public awareness campaigns are a cornerstone in building public trust, promoting transparency, and ensuring the success and acceptability of deer management initiatives within the community. Through these campaigns, a conducive environment for effective and sustainable deer management can be fostered, ensuring the enduring vibrancy and richness of natural landscapes.

COMMUNITY-BASED DEER MANAGEMENT PROGRAMS

Community-based deer management programs embody a collaborative approach wherein local communities play a significant role in managing deer populations. These programs are predicated on the notion that involving local stakeholders in decision-making and implementation processes engenders a sense of ownership and enhances the success and sustainability of deer management initiatives. They encapsulate a blend of local knowledge, shared responsibility, and cooperative action, creating a fertile ground for successful deer management.

Engagement and Participation: Community engagement and participation are the linchpins of community-based deer management programs. By involving local residents, landowners, and other stakeholders in planning and decision-making processes, these programs foster a sense of communal ownership and responsibility. This engagement not only enriches the management program with local knowledge and insights but also creates a network of stakeholders invested in the program's success.

Capacity Building and Education: A significant aspect of community-based deer management programs is the emphasis on capacity building and education. Through training, workshops, and educational campaigns, community members are equipped with the requisite knowledge and skills to participate actively in deer management activities. This capacity building is instrumental in empowering local communities, enabling them to take informed actions, and contribute effectively to deer management initiatives.

Leveraging Local Knowledge: Local communities often possess a wealth of knowledge about deer behaviour, local habitats, and existing challenges. Community-based deer management programs leverage this local knowledge to design and implement more effective and context-specific management strategies. The amalgamation of local insights with scientific knowledge creates a robust foundation for devising strategies that are both effective and locally acceptable.

Shared Responsibility and Resource Pooling: Community-based programs engender a sense of shared responsibility among local stakeholders. This shared responsibility often translates into resource pooling, where community members contribute resources, time, and expertise towards deer management activities. Resource pooling is a pragmatic way to augment the resources available for deer management, thereby enhancing the scope and impact of management initiatives.

Conflict Resolution and Community Consensus: Community-based approaches provide a platform for dialogue, conflict resolution, and building community consensus on deer management strategies. Through open discussions and participatory decision-making, conflicts can be addressed, and a consensus can be built on the way forward. This consensus-building is crucial for creating a harmonious environment conducive to successful deer management.

Feedback and Adaptive Management: Community-based deer management programs often have built-in mechanisms for feedback and adaptive management. By continuously engaging with the community, receiving feedback, and adapting strategies accordingly, these programs ensure that deer management initiatives remain relevant, effective, and aligned with community needs and expectations.

Economic Benefits and Livelihoods: Community-based deer management can also have economic dimensions, providing avenues for local economic development and livelihoods. For instance, community-based venison processing and marketing initiatives can generate local employment and income, thereby contributing to local economic sustainability.

Community-based deer management programs epitomize a cooperative approach that merges local engagement with scientific management principles. Through these programs, deer management transcends the conventional boundaries, morphing into a community-driven endeavour that holds promise for sustainable and effective deer management. By fostering a culture of cooperation, shared responsibility, and continuous learning, community-based deer management programs pave the way for a harmonious co-existence between human communities and deer populations, ensuring the enduring vibrancy and ecological balance of our natural landscapes.

TRAINING AND CERTIFICATION

Training and certification are integral components of ensuring that deer management practices are executed proficiently and ethically. These programs serve to equip individuals and entities involved in deer management with the requisite knowledge, skills, and credentials necessary for effective and humane management of deer populations.

Enhancing Competence: Training programs in deer management are designed to instil a high level of competence among practitioners. Through a structured curriculum that encompasses both theoretical knowledge and practical skills, individuals are prepared to handle the myriad challenges associated with managing deer populations. This includes understanding deer biology and behaviour, mastering population monitoring techniques, and learning effective and humane population control strategies.

Standardizing Practices: Certification programs serve to standardize deer management practices across the board. By establishing a set of standards and benchmarks, certification programs ensure that individuals and entities engaged in deer management adhere to accepted best practices and regulatory requirements. This standardization is pivotal in maintaining a high level of professionalism and consistency in deer management practices.

Ensuring Ethical Conduct: Ethics are at the core of deer management practices. Training and certification programs often have a strong emphasis on ethical conduct, ensuring that deer management is carried out humanely and in accordance with established welfare standards. By instilling a strong ethical foundation, these programs promote responsible and ethical deer management practices that prioritize the welfare of deer and the integrity of ecosystems.

Promoting Safety: Deer management activities, particularly population control measures such as culling, entail certain risks. Training programs equip practitioners with the knowledge and skills necessary to carry out these activities safely. By promoting a culture of safety and adherence to safety protocols, training and certification contribute significantly to minimizing risks associated with deer management activities.

Fostering Public Trust: Public trust is a crucial aspect of successful deer management initiatives. When deer management activities are carried out by trained and certified individuals, it engenders a higher level of trust among the public. This trust is instrumental in garnering support for deer management initiatives and promoting a culture of cooperation and compliance among stakeholders.

Facilitating Regulatory Compliance: Compliance with regulatory requirements is a fundamental aspect of deer management. Training and certification programs provide practitioners with a thorough understanding of the legal and regulatory landscape surrounding deer management. This understanding is crucial for ensuring compliance with laws, regulations, and guidelines governing deer management activities.

Continuous Learning and Professional Development: The field of deer management is dynamic, with new challenges and solutions emerging continually. Training and certification programs often provide avenues for continuous learning and professional development. Through ongoing training, practitioners can stay abreast of the latest developments, best practices, and emerging technologies in deer management.

Accrediting Professionalism: Certification programs provide a platform for accrediting professionalism in deer management. By earning certification, individuals and entities demonstrate their commitment to maintaining a high level of professionalism, competence, and ethical conduct in their deer management practices.

The synergy of training and certification in deer management fosters a conducive environment for effective, ethical, and professional deer management practices. Through these programs, the deer management sector is poised to achieve a high level of professionalism and effectiveness, ensuring that deer populations are managed in a humane, sustainable, and socially acceptable manner.

CHAPTER ELEVEN

POLICY, LEGISLATION, AND COMPLIANCE

UK WILDLIFE LAWS AND REGULATION

Please note these are constantly changing therefore you should reach out to the relevant department for updated details.

The legal framework governing wildlife management in the UK is comprehensive, with specific laws and regulations that pertain to deer management. This legal canvas seeks to balance the conservation and management of deer populations with the interests of agriculture, forestry, and public safety.

Legislative Framework: The legislative framework for deer management in the UK is primarily enshrined in various acts and regulations. Key among them is the Deer Act 1991 and the Wildlife and Countryside Act 1981, which provide the legal foundation for deer management. These acts delineate the rights, responsibilities, and procedures for managing deer populations, including provisions for culling, fencing, and other management interventions.

Regulatory Authorities: The enforcement and administration of wildlife laws, including those pertaining to deer management, are vested in various regulatory authorities. In the UK, the primary authorities include the Department for Environment, Food & Rural Affairs (DEFRA) and its devolved counterparts in Scotland, Wales, and Northern Ireland. Additionally, bodies like Natural England, Scottish Natural Heritage, and the Forestry Commission play significant roles in regulating and supporting deer management activities.

Licensing and Permits: The legal framework provides for a system of licensing and permits that regulate deer management activities. This licensing framework ensures that such activities are carried out by qualified and competent individuals or entities, and in a manner that conforms to the stipulated legal and ethical standards.

Seasonal Restrictions and Close Seasons: The law stipulates seasonal restrictions on certain deer management activities to ensure the welfare of deer populations, particularly during sensitive periods such as breeding seasons. These seasonal restrictions, known as close seasons, are crucial for promoting ethical and humane deer management practices.

Note: Changes to deer management legislation. The change to the Close Season Order removed the close season for all species of male deer in Scotland effective from 21 October 2023. You are no longer required to apply to NatureScot for an Out of Season Authorisation to shoot male deer.

Landowner Rights and Responsibilities: Landowners have specific rights and responsibilities under the law concerning deer management on their property. They are often permitted to undertake certain management activities to protect their land and interests, provided they adhere to the legal requirements and obtain the necessary permits or licenses.

The UK's legal framework for deer management is structured

to ensure that deer populations are managed in a sustainable, ethical, and socially acceptable manner. By delineating the rights, responsibilities, and procedures for deer management, the law provides a structured approach to balancing the diverse interests and challenges associated with deer management. Through this legal canvas, a conducive environment for effective and responsible deer management is fostered, ensuring the harmonious coexistence of deer populations with human communities and the natural environment.

REPORTING AND DOCUMENTATION

Accurate reporting and documentation are foundational to effective deer management, creating a structured pathway towards a thorough understanding and proficient control of deer populations. The essence of these processes transcends mere statutory requisites, embedding itself within the core necessity for a methodical approach to management strategies.

Record-Keeping for Informed Decision Making: The accurate recording of data pertaining to deer populations, their health, and the impact of management strategies is instrumental in forming the basis for informed decision-making. Comprehensive records furnish a clear portrayal of population trends, the efficacy of various management interventions, and areas warranting attention. This data is indispensable for the assessment of current management practices and the blueprinting of future strategies. For instance, data on population dynamics over a period can guide crucial decisions regarding culling or habitat restoration initiatives, ensuring they are based on solid empirical evidence rather than conjecture.

Legal Compliance and Audit Trails: In the UK, the governance of deer management is encapsulated within various legislative frameworks. Adherence to accurate reporting and documentation is quintessential for compliance with these legal frameworks, providing an auditable trail for regulatory scrutiny. Documentation of actions undertaken, such, as culling or relocation efforts, and their outcomes are paramount for demonstrating compliance with legal and ethical standards, especially when introducing venison into the food chain. Furthermore, in the event of disputes or legal challenges, well-maintained records furnish the necessary evidentiary support to validate the actions executed by deer management authorities.

Transparency and Public Engagement: The fostering of transparency in deer management operations is pivotal for engendering public trust and support. Accurate reporting and documentation are conduits to transparency by availing factual information regarding deer management activities. When disseminated to the public, this information aids in crafting an informed community, a critical element for the successful orchestration of deer management. The engagement of the public with accurate information can also cultivate a collaborative ethos towards deer management, where the community can actively partake in management initiatives, thus fostering a communal sense of ownership and responsibility.

Resource Allocation and Funding: The essence of accurate reporting is seminal for effective resource allocation. By scrutinising the data collated, resources can be channelled towards areas of paramount need. Additionally, meticulously maintained documentation often serves as a prerequisite for securing funding and grants for deer management programmes. Funding entities necessitate clear records demonstrating the need for funds and the envisioned utilisation of these resources, ensuring they are being allocated towards a structured and well-planned initiative.

For example, those claiming the Countryside Stewardship Grant WS1 in 2023, must record the following:

- A Forestry Commission approved management plan that justifies the need for this option
- A Deer Management Plan in place by the end of the first year
- Monitoring reports for year 1, 3 and 5 of the agreement to confirm progress (for example providing before and after photographs, a record of the number of deer culled, and the results of squirrel monitoring)
- Evidence of activities undertaken through monitoring, photography and marking
- Any bank statements, receipted invoices, consents, or permissions connected with the work
- Records of all management activity on the option area for each parcel, including an operational site assessment (or similar) to show UKFS compliant operational activities

Capacity Building and Training: The data amassed and documented over time burgeons as a valuable asset for training and capacity-building endeavours. It proffers real-world insight into deer management challenges and the effectiveness of different strategies, which can be utilised to educate and train new personnel transitioning into the field of deer management. This treasure trove of information serves as a living repository of knowledge, aiding in the continuous enhancement of skills and understanding amongst deer management professionals.

The Ethical Steward

CHAPTER TWELVE

CASE STUDIES AND BEST PRACTICES

HYPOTHETICAL DEER MANAGEMENT PROGRAMS

Though the field is rich with real-world examples, envisioning hypothetical scenarios can also provide a wealth of understanding. These imagined scenarios can offer a sandbox for exploring different strategies, challenges, and outcomes in a variety of contexts. Below, we delve into a series of hypothetical deer management programs, each crafted to address specific challenges and to showcase a range of management strategies.

Program Alpha: Integrated Forest Management: Program Alpha envisages a scenario in a large forested region grappling with overpopulation of deer and resultant habitat degradation. The cornerstone of this program is integrating deer management by means of targeted culls and fencing within the broader ambit of forest management to nurture biodiversity, and bolster overall ecosystem health. Through this approach, the program underscores the imperative of a holistic perspective, recognizing the intertwined destinies of deer populations and their habitat.

Program Beta: Urban Deer Management: Urban areas are becoming increasingly entangled with the lives of deer, leading to unique management challenges. Program Beta is tailored to address these urban deer issues employing a spectrum of lethal and non-lethal measures such as public education, highway hotspot warning, fencing and targeted culling. At its core, the program highlights the vital role of community involvement and continuous education. It combines these elements as the linchpins of garnering local support, harnessing local knowledge, and fostering a respectful co-existence between humans and deer in urban landscapes.

Program Gamma: Agricultural Deer Management: In regions where agriculture forms the economic backbone, Program Gamma aims to shield the agricultural lands from deer-induced damage. It employs a medley of deterrents, fencing, controlled culling, and explores compensation schemes for beleaguered farmers. The program underscores the necessity of adaptive strategies, tuned to the unique circumstances of different agricultural settings. It also throws light on the quest for cost-effective solutions, a quest that is quintessential for upholding the economic vigour of agricultural communities.

Program Delta: Collaborative Deer Management: Program Delta sketches a realm where multiple landowners and stakeholders rally together to address deer overpopulation and associated challenges. It endeavours to pool resources and expertise in a collective pursuit of shared deer management goals. The narrative of Program Delta accentuates the essence of stakeholder engagement and shared resources. It unfolds a collective action, showcasing how united endeavours can unfurl more efficient and cost-effective deer management initiatives.

Search 'Sussex Woods Protected Sites Pilot Scheme 2023' for an example.

These hypothetical scenarios, though not drawn from real-world experiences, furnish a canvas to explore a myriad of strategies and considerations pivotal to deer management across diverse contexts. They serve as conceptual frameworks, each sketching a different set of circumstances, challenges, and approaches. Through the contemplation of these scenarios, deer managers and stakeholders can garner a broader understanding and envision a variety of pathways to navigate the complex landscape of deer management.

CHAPTER THIRTEEN

RESOURCES AND SUPPORT

RELEVANT ORGANIZATIONS AND AGENCIES

In the UK, a number of organizations and agencies play pivotal roles in deer management, either by providing advice, training, or setting standards for best practices.

Here's a brief overview of some notable entities involved in this domain:

British Association for Shooting and Conservation (BASC): The BASC deer team promotes sustainable deer management across the UK by offering advice on habitat and species management, developing and supporting best practices, providing deer stalking opportunities, and influencing government policy to protect shooting interests.

The British Deer Society (BDS): BDS operates throughout the UK with a focus on educating and raising awareness about deer and related issues. They also provide training to ensure deer are managed humanely and to the highest standards. The society is committed to inspiring individuals about deer while also engaging in education and research.

Deer Management Groups (DMGs): DMGs are crucial where

the natural range of wild deer extends across areas of multiple land ownership. They act as forums for addressing local deer management issues and promote collaborative management efforts to tackle challenges posed by wild deer populations.

National Quality Assurance Scheme for Wild Venison: This scheme is a cross-sector initiative developed by several entities including the Forestry Commission, Grown in Britain, Forestry England, Natural Resources Wales, and the National Game Dealers Association. It's aimed at ensuring the quality of wild venison, which indirectly plays a role in deer management by providing a controlled and sustainable outlet for deer populations.

These organizations and schemes provide a structured framework for deer management in the UK, catering to various aspects from habitat conservation to public education and quality assurance in venison production. Through their distinct yet complementary roles, they contribute to a comprehensive approach towards sustainable deer management.

CHAPTER FOURTEEN

FUTURE TRENDS AND CHALLENGES

EMERGING ISSUES IN DEER MANAGEMENT

Deer management is a field that constantly evolves in response to the changes of ecological, social, and technological factors. As new challenges arise, so do opportunities for innovation and enhancement in management practices. Here we discuss some of the emerging issues in deer management, shedding light on the pathway of challenges and prospective advancements that lay ahead.

Urbanisation: The encroachment of urban areas into traditional deer habitats is a burgeoning challenge. Urban deer populations can cause numerous problems including traffic accidents, damage to vegetation, and the potential for disease transmission. Conversely, urban areas can also provide refuges for deer from hunting and predation, leading to overpopulation and consequent challenges in population control.

Technological Advancements: The advent of new technologies offers a suite of tools for better deer management. Advances in GPS tracking, drone technology, and data analytics can provide more accurate population estimates, monitor deer movement, and assess the impact of deer on habitats. Social media and mobile applications can facilitate public engagement and education, fostering a collaborative approach to deer management.

Emerging Diseases: The emergence and spread of new diseases pose significant threats to deer populations and can have cascading effects on ecosystems. Effective monitoring, rapid response, and public education are essential to mitigate the impacts of emerging diseases and ensure the health and welfare of deer populations.

Public Perceptions and Social Acceptance: Public attitudes towards deer management, particularly lethal control measures like culling, are evolving. Achieving social acceptance and understanding of the necessity for certain management practices is crucial. Engaging the public in decision-making processes and educating them about the science and ethics of deer management can foster broader support for necessary management actions.

Innovative Management Strategies: The development of innovative management strategies, such as integrated wildlife management plans, presents new opportunities for humane and effective deer management. Collaborative initiatives that bring together landowners, managers, policymakers, and the public can yield holistic solutions to the complex challenges posed by deer management.

The pathway of deer management is laden with both challenges and opportunities. Navigating this pathway requires a vigilant eye on emerging issues, a willingness to adapt, and a commitment to fostering a dialogue among all stakeholders involved. As we move forward, the integration of technological innovations, public engagement, and adaptive management strategies will be pivotal in addressing the emerging issues in deer management, ensuring a co-existence between humans, deer, and the environment.

RECOMMENDATIONS FOR POLICY AND PRACTICE

Navigating the complex terrain of deer management necessitates a coherent blend of policy initiatives and practical interventions. Here we provide a constellation of recommendations aimed at fortifying the efficacy, ethical underpinning, and societal acceptance of deer management strategies.

Adaptive Management Framework: Employing an adaptive management framework can foster a more responsive and effective deer management approach. This framework allows for the continuous monitoring of deer populations and their impacts, with the flexibility to adjust management strategies based on the evolving circumstances and the outcomes of previous interventions.

Technological Integration: Integrating modern technologies like GPS tracking, drones, and data analytics into deer management practices can significantly enhance monitoring, data collection, and public engagement. Leveraging these technologies can provide more accurate information, enabling informed decision-making and effective management interventions.

Collaborative Initiatives: Fostering collaborative initiatives among landowners, government agencies, NGOs, research institutions, and the public can yield holistic solutions. Cross-sectorial collaborations can facilitate the sharing of knowledge, resources, and expertise, thereby bolstering the collective capability to address deer management challenges.

Investment in Research: Continual investment in research to understand deer ecology, behaviour, and the impacts of deer-human interactions is imperative. Research can unearth novel insights, improve management strategies, and provide evidence-based recommendations for policy and practice.

Policy Harmonization: Ensuring that policies governing deer management are harmonized across different jurisdictions and are aligned with contemporary scientific understanding and societal values is crucial. A coherent policy framework can provide clear guidelines, promote legal compliance, and ensure that deer management practices are ethical and socially just.

Training and Capacity Building: Implementing training and certification programs for deer management professionals can promote a high standard of professionalism, ethics, and competency. Continuous professional development opportunities can ensure that practitioners are well-equipped with the latest knowledge and skills required for effective deer management.

Economic Incentives: Exploring economic incentives such as grants, subsidies, or revenue-generating activities like venison production can provide the financial support necessary for sustainable deer management. Economic incentives can also encourage landowners and local communities to engage actively in deer management initiatives.

These recommendations embody a multidimensional approach aimed at navigating the future landscape of deer management. Through a blend of adaptive management, technological integration, public engagement, collaborative initiatives, research investment, policy harmonization, capacity building, and economic incentives, we can aspire to address the impending challenges and leverage the emerging opportunities in deer management, thus striving towards a harmonious co-existence between deer, humans, and the ecosystem at large.

CHAPTER FIFTEEN

CONTACT INFORMATION

CONTACT DETAILS FOR RELEVANT AUTHORITIES AND ORGANIZATIONS:

Below are the contact details for relevant authorities and organizations that play a pivotal role in deer management within the UK. It's imperative to note that the contact information provided may change over time, and it's advisable to verify the details through official channels.

Government Authorities:

Natural England
Website: www.gov.uk/government/organisations/natural-england
General Enquiries: 0300 060 3900

Scottish Natural Heritage
Website: www.nature.scot
General Enquiries: 01463 725000

Natural Resources Wales
Website: www.naturalresources.wales
General Enquiries: 0300 065 3000

Northern Ireland Environment Agency
Website: www.daera-ni.gov.uk/niea
General Enquiries: 028 9056 9568

Deer Management Organizations:

The Deer Initiative
Website: www.thedeerinitiative.co.uk
General Enquiries: 01829 730 969

British Deer Society
Website: www.bds.org.uk
General Enquiries: 01425 655434

Research Institutions:

Game & Wildlife Conservation Trust
Website: www.gwct.org.uk
General Enquiries: 01425 651000

Mammal Society
Website: www.mammal.org.uk
General Enquiries: 02380 010981

Engaging with these organizations and authorities can provide invaluable insights, support, and guidance in deer management endeavours. Whether seeking advice, technical support, or collaborative opportunities, these entities form a robust network of expertise and resources crucial for fostering effective and sustainable deer management practices.

REFLECT, APPLY, AND EXPLORE

Thank you for delving into the pages of this guide on deer management. As you reach the end, the journey of exploration and application begins. The knowledge acquired herein is a seed; its true potential unfolds as you apply it to your real-world challenges in deer management.

Deer management is a domain where experience is a profound teacher. Each parcel of land, each herd of deer, brings with it unique challenges and lessons. By applying the principles discussed in this guide, you are well on your way to becoming not just a manager of deer, but a steward of the land.

The landscape of deer management is ever-evolving with new research, methodologies, and technologies emerging. While this guide provides a solid foundation, continuous learning and adaptation are crucial. Engage with local and online communities of deer managers, attend workshops, and stay updated with the latest in deer management research and practices.

Furthermore, consider sharing your experiences and insights with others in the field. The tradition of mentorship and knowledge sharing is a hallmark of a thriving deer management community. Your experiences, whether they entail triumphs or challenges, are valuable lessons for others.

Lastly, should you have any inquiries, feedback, or simply wish to share your deer management journey, feel free to reach out. Your insights and experiences are not only a reflection of your growth but a contribution to the broader deer management community.

As you close this guide, may your passion for deer management continue to grow, leading to prosperous lands, healthy deer populations, and a thriving community of ethical and knowledgeable deer managers.